YOUR KNOWLEDGE HAS VALUE

- We will publish your bachelor's and
  master's thesis, essays and papers

- Your own eBook and book -
  sold worldwide in all relevant shops

- Earn money with each sale

Upload your text at www.GRIN.com
and publish for free

Imprint:

Copyright © 2015 GRIN Verlag, Open Publishing GmbH
Print and binding: Books on Demand GmbH, Norderstedt Germany
ISBN: 9783668484276

This book at GRIN:

http://www.grin.com/en/e-book/370809/collocation-method-for-weakly-singular-
volterra-integral-equations-of-the

Henry Ekah-Kunde

Collocation method for Weakly Singular Volterra Integral Equations of the Second Type

GRIN Publishing

Collocation method for weakly singular Volterra integral equations of the second type

Henry Eneme Ekah-Kunde

School of Science, Enginering and Technology

Saint Monica University, Buea

Abstract: *Volterra integral equations of weakly singular types have solutions which are non-smooth near the initial point of the integration interval. The implementation of the collocation spline method will lead us to the examination of the attainable order of convergence of this method on graded mesh points for non linear Volterra integral equations with singular kernels.*

Keywords: Weakly singular Volterra integral equation; Numerical method: Collocation method

Content

1. Introduction

In scientific and engineering problems Volterra integral equations are always encountered. Applications of Volterra integral equations arise in areas such as population dynamics, spread of epidemics in the society, heat and fluid conductivity, etc. Some classes of initial boundary problems of heat conductivity have been transformed to an equivalent system of Volterra type integral equations of second type (Jalalvand, Jazbi, & Mokhtarzadeh, 2013). The problem statement is to obtain a good numerical solution to such an integral equation.

Numerical solutions of second kind Volterra integral equations with weakly singular kernels have been investigated by Brunner (1985), Brunner, Pedas & Vainikko (1997), Brunner and van der Houwen (1986), Linz (1985), Makroglou (1981), Te Riele (1982), Diogo and Lima (2007).

In this paper we shall present a brief theory of Volterra Integral equation, particularly, of weakly singular types. We are interested in finding an approximate solution which exhibits high order of convergence. This will lead us to the implementation of an approximation with polynomial splines on special graded meshes. This implementation will be carried out along the lines of Brunner (1985), Brunner and van der Houwen (1986), Te Riele (1982) and Brunner, Pedas & Vainikko (1997). The principle of this method is to approximate the exact solution of the equation in a suitable finite dimensional space.

2. Volterra integral equations with weakly singular kernel

An integral equation is defined to be a functional equation, whereby the unknown function features under the integral sign. A Volterra integral equation, an equation where the unknown function appears under the integral sign, is characterized by a variable upper limit and a fixed lower limit of integration. Volterra integral equations are classified in three main groups: first, second and third kind. In this paper we shall be dealing with (nonlinear) Volterra integral equation of second kind of the form:

$$x(t) = c(t) + \int_0^t k(t,s,x(s))\,ds,\tag{2.1}$$

with $x : [0,T] \to R$ being the unknown function, $c: [0,T] \to R$ is known as the initial function, and $k(t,s,x)$ continuous on $[0,T] \times [0,T] \times R$ is called the kernel; with $0 \le s \le T, 0 < T < \infty$.

Equation (2.1) is said to be linear if its kernel has the form

$$k(t,s,x) = K(t,s)x.$$

A (nonlinear) Volterra integral equation of second kind with weakly singular kernel is the equation of type (2.1) in which the kernel $k(t,s,x)$ is replaced by

$$k(t,s)b(t,s,x),$$

i.e. the product of a smooth function and a weakly singular (unbounded but integrable) function. $k(t,s,x)$ is defined in the following domain: $k: [0,T] \times [0,T] \setminus \{(t,s): s \ge t\} \to R)$.

The kernel $k(t,s,x)$ is assumed to satisfy the following conditions:

$|k(t,s)| \le a(t-s)^{-\alpha}$, with a > 0 and $\alpha \in (0,1)$. Hence the kernel is singular at $t = s$. An example of such a kernel is $k(t,s) = (t-s)^{-\alpha}$ with $0 < \alpha < 1$.

In the rest of this paper we shall be dealing with nonlinear weakly Volterra integral equations of second kind of the form

$$x(t) = c(t) + \int_0^t (t-s)^{-\alpha} b(t,s,x(s))\,ds.\tag{2.2}$$

2.1 Existence and uniqueness of the solution

Let K denote the Volterra integral operator on $C[0,T]$, and defined as follows:

$$(Kx)(t) = c(t) + \int_0^t k(t,s,x(s))\, \mathrm{d}s. \qquad\qquad (2.3)$$

Definition 2.1: *Let K be a Banach space. The point $x \in X$ is called a fixed point of the operator K if $Kx = x$.*

Lemma 2.1 *[Contraction mapping theorem]: Let X be a closed subset of a Banach space, and let the operator $K : X \to X$. Suppose there exists a number $\delta < 1$ such that for all $x, y \in X$*

$$\|Kx - Ky\| \leq \delta\|x - y\|.$$

Then K has a unique fixed point in X, (Gohberg and Goldberg, 1981, p. 256).

Theorem 2.1 *(Existence and uniqueness of the solution):*

Let $c \in C[0,T]$ and $k(t,s,x)$ be continuous in $[0,T] \times [0,T] \times R$. Let $k(t,s,x)$ also satisfy a Lipschitz condition with respect to x, that is

$$\|k(t,s,x_1) - k(t,s,x_2)\| \leq c_L|x_1 - x_2|$$

$$\forall\ (t,s) \in [0,T] \times [0,T]\ \text{and}\ x_1, x_2 \in R,$$

with c_L known as the Lipschitz constant. Then the nonlinear Volterra integral equation (2.1) possesses a unique solution $x \in C[0,T]$.

This theorem is a carry-over from Linz (1985) pp 52-53. The theorem can also be extended to our specific Volterra integral equation with weakly singular kernel (2.2) as follows:

Theorem 2.2: *Suppose that $c \in C[0,T]$ and that $b(t,s,x)$ is continuous in $[0,T] \times [0,T] \times R$, as well as $k(t,s)$ is weakly singular. Let $b(t,s,x)$ also satisfy a Lipschitz condition with respect to x with Lipschitz constant c_L. Then the nonlinear Volterra integral equation (2.2) possesses a unique solution $x \in C[0,T]$.*

3. Numerical Method – the Collocation method

In this section we shall handle a numerical method – the collocation method – for solving weakly singular integral equations of the form (2.2). In the collocation method we approximate the exact solution of a given functional equation in a suitable chosen finite subset of the interval on which the equation is to be solved. The approximating spaces applied to this problem are certain polynomial spline spaces.

3.1 The approximating spline spaces

The weakly singular Volterra integral equation (2.2) will be solved in the interval $I := [0,T]$.

For a given $N \in \mathbb{N}$ let $\Pi_N := \{t_0, t_1, \dots, t_N\}$ with $(0 = t_0 < t_1 < \dots < t_N = T)$ be the mesh of the interval I.

$$Z_N := t_k : 1 \leq k \leq N - 1,$$

denotes the set of the interior points of the partition Π_N, and associates as well the subinterval $\sigma_0 := [t_0, t_1]$, $\sigma_k := (t_k, t_{k+1}]$ for $k = 1, \dots, N-1$.

Definition 3.1: *Let m and d be given integers satisfying $-1 \leq d \leq m - 1$. With γ_m we denote the space of real polynomials of degree not greater than m.*

$$S_m^d(Z_N) := \{ y : y|_{\sigma_k} =: y_k \in \gamma_m \quad k = 0, \dots, N\text{-}1 \} \text{ with } d \in \{0,1\}$$

is called the space of real polynomial splines of degree m and continuity class d with knots Z_N.

The space S_m^d

- has finite dimension, given by $\dim S_m^d = N(m - d) + d + 1$,

- is linear, that is, suppose α and β are real constants and x, y are elements of this space, then $\alpha x + \beta y$ is also an element of the same space.

We define the collocation set by:

$$B(N) := \bigcup_{k=0}^{N-1} B_k,$$

where $B_k := \{t = t_{kj} := t_k + c_j h_k; j = 1, \ldots m; k = 0, \ldots N-1\}$ with the real numbers

$0 \leq c_1 < c_2 < \ldots c_m \leq m$ known as the collocation parameters, and $h_k = t_{k+1} - t_k$ is known as the diameter of the kth $-$ mesh. The unique solution of (2.2) will be approximated by an element $y \in S_{m-1}^{(-1)}$, whereby the elements of $S_{m-1}^{(-1)}(Z_N)$ may possess at most N-1 discontinuities at the knots Z_N. In other words the elements of $S_{m-1}^{(-1)}(Z_N)$ are not continuous in the entire interval $[0,T]$. In the case where one chooses $c_1 = 0$ and $c_m = 1$, the approximating element will be an element of the smoother polynomial spline space $S_{m-1}^{(0)}(Z_N)$ and continuous on the whole interval $[0,T]$ (Brunner (1985), p. 419).

In order to solve equation (2.2) we therefore project x into the space $S_{m-1}^{(-1)}$, that is, we determine $y \in S_{m-1}^{(-1)}$ such that

$$y_k(t) = c(t) + \int_0^t (t-s)^{-\alpha} b(t,s,y(s))ds \tag{3.1}$$

is satisfied for all $t \in B_k$, $k = 0, \ldots, N-1$.

3.2 Discretization of the collocation equation

We start by rewriting the weakly singular Volterra integral equation (2.2) for $t \in \sigma_k$ in "one-step form" as follows:

$$x(t) = c(t) + \sum_{i=0}^{k-1} \int_{t_i}^{t_{i+1}} (t-s)^{-\alpha} b(t,s,y(s))ds + \int_{t_i}^t (t-s)^{-\alpha} b(t,s,y(s))ds \quad \text{for } k = 0, \ldots N-1. \tag{3.2}$$

Analogous to equation (3.1), that is approximating the function $x(.)$ with $y \in S_{m-1}^{(-1)}(Z_N)$, equation (3.2) reads, for $t = t_{kj}$,

$$y_k(t_{kj}) = c(t_{kj}) + \sum_{i=0}^{k-1} \int_{t_i}^{t_{i+1}} (t_{kj} - s)^{-\alpha} b(t_{kj}, s, y_i(s)) ds + \int_{t_k}^{t_{kj}} (t_{kj} - s)^{-\alpha} b(t_{kj}, s, y_k(s)) ds,$$

where $t_{kj} = t_k + c_j h_k$ with c_j and h_k defined as above. Substituting $s = t_i + vh_i$ and $s = t_k + vh_k$ in

the respective intervals, and setting these in the last equation, we obtain

$$y_k(t_{kj}) = c(t_{kj}) + \sum_{i=0}^{k-1} h_i \int_0^1 (t_{kj} - t_i - vh_i)^{-\alpha} b(t_{kj}, t_i + vh_i, y_i(t_i + vh_i)) dv$$

$$+ h_k^{1-\alpha} \int_0^{c_j} (c_j - v)^{-\alpha} b(t_{kj}, t_k + vh_k, y_k(t_k + vh_k)) dv, \tag{3.3}$$

for $j = 1, \dots, m$; and $k = 0, 1, \dots, N - 1$; whereby $y_k(t_{kj}) \in \gamma_{m-1}$. After obtaining the values

$y_k(t_{kj})$ for $j = 1, \dots, m$, we then acquire from equation (3.3) the following approximation

$$y_k(t_k + vh_k) = \sum_{l=1}^m L_l(v) y_k(t_{kl}) \quad \text{with } (t_k + vh_k) \in \sigma_k \text{ for } k = 0, 1, \dots, N - 1, \tag{3.4}$$

whereby

$$L_l(v) := \prod_{k=1, k \neq l}^m \frac{v - c_k}{c_l - c_k} \quad \text{with } (l = 1, \dots, m) \tag{3.5}$$

denotes the lth Lagrange fundamental polynomial associated with the collocation parameters c_l.

Equation (3.3) can be written in a compact form as follows:

$$y_k(t_{kj}) = c(t_{kj}) + h_k^{1-\alpha} \Phi_{kk}^{(j)}[y_k] + \sum_{i=0}^{k-1} h_i^{1-\alpha} \Phi_{ki}^{(j)}[y_i], \quad \text{with } j = 1, 2, \dots, m; k = 0, 1, \dots, N-1, \tag{3.6}$$

whereby

$$\Phi_{ki}^{(j)}[y_i] := \int_0^1 \left(\frac{t_{kj} - t_i}{h_i} - v \right)^{-\alpha} b\big(t_{kj}, t_i + vh_i, y_i(t_i + vh_i)\big) dv \quad \text{for } i = 0, 1, \dots, k-1,$$

$$\Phi_{kk}^{(j)}[y_k] := \int_0^{c_j} (c_i - v)^{-\alpha} b\left(t_{kj}, t_k + vh_k, y_k\left(t_k + vh_k\right)\right) dv.$$

The integrals in equation (3.3) can, usually, not be evaluated analytically. In order to obtain the values of these integrals with more precision further discretization will be necessary. The full discretization of (3.3) is obtained by approximating these integrals with the help of suitable product integration techniques (interpolatory m-point product quadrature formulas, see Brunner and van der Houwen (1986), Chapter 2, pp. 364, 375). We denote by $\hat{\Phi}_{ki}^{(j)}[\hat{y}_i]$ and $\hat{\Phi}_{kk}^{(j)}[\hat{y}_k]$ the approximation to $\Phi_{ki}^{(j)}[y_i]$ and $\Phi_{kk}^{(j)}[y_k]$ respectively:

$$\hat{\Phi}_{ki}^{(j)}[\hat{y}_i] := \sum_{l=1}^{m} w_{jl}^{(k,i)}(\alpha) b\left(t_{kj}, t_i + d_l h_i, \hat{y}_l\left(t_i + d_l h_i\right)\right),$$

$$\hat{\Phi}_{kk}^{(j)}[\hat{y}_k] := \sum_{l=1}^{m} w_{jl}(\alpha) b\left(t_{kj}, t_k + d_{jl} h_k, \hat{y}_l\left(t_k + d_{jl} h_k\right)\right), \tag{3.7}$$

with $0 \leq d_1 <, ..., < d_{\mu 1} \leq 1;$ $0 \leq d_{j,1} <, ..., < d_{j,\mu 0} \leq c_j$ for $(j = 1, 2, ..., m)$.

Hereby the $d_1 <, ..., < d_{\mu 1}$ are the quadrature abscissas in the interval $[0,1]$, while the $d_{j,1} <, ..., < d_{j,\mu 0}$ are those in the integration interval $[0, c_j]$. The weights of these integrals are given as follows:

$$w_{jl}^{(k,i)}(\alpha) = \int_0^1 \left(\frac{t_{kj} - t_i}{h_i} - v\right)^{-\alpha} \lambda_l(v) dv,$$

$$w_{jl}(\alpha) = \int_0^{c_j} (c_i - v)^{-\alpha} \lambda_{jl}(v) dv;$$

with $\lambda_l(v) = \prod_{r=1}^{m}\left(\frac{v - d_r}{d_l - d_r}\right),$ $\lambda_{jl}(v) = \prod_{r=1}^{m}\left(\frac{v - d_{jr}}{d_{jl} - d_{jr}}\right),$ with $r \neq l.$

The λ_l, λ_{jl} describe Lagrange fundamental polynomials, which are associated with the quadrature parameters d_r, d_{jr}, respectively. After successful application of the quadrature approximation to our collocation equation, we finally obtain the fully discretized version of (3.6) as follows:

$$\hat{y}_k(t_{kj}) = c(t_{kj}) + h_k^{1-\alpha}\Phi_{kk}^{(j)}[\hat{y}_k] + \sum_{i=0}^{k-1} h_i^{1-\alpha}\Phi_{ki}^{(j)}[\hat{y}_i]. \tag{3.8}$$

Hereby is $\hat{y}$ an element of the polynomial spline space $S_{m-1}^{(-1)}(Z_N)$. It is an approximation of and slightly different from $y \in S_{m-1}^{(-1)}(Z_N)$ due to errors induced by the product integration process.

Remarks: The product integration technique was implemented in such a way that evaluation of the weakly singular kernel was not carried out for $s > t$, that is, in the domain, where the kernel is not defined. Therefore the choice of the quadrature abscissas in the integration interval $[0, c_j]$ was appropriately made so as to maintain the weakly singularity of our kernel. In other words $(c_i - v)^{-\alpha} \neq \infty$, because in the interval $[t_k, \; t_k + c_j h_k]$ the interpolation abscissas will be the points $\{t_k + c_j c_l h_k : l = 1, \ldots, m\}$.

It is also generally acceptable to set $d_l = c_l,\; d_j d_l = c_j c_l;\; \mu_0 = \mu_1 = m,\; j = 1, \ldots, m$. Therefore we shall use, henceforth, c_l in place of the d_l, and also substitute $\lambda_l(v)$ with $L_l(v)$.

In every subinterval $\sigma_k := (t_k, t_{k+1}]$ we obtain $\hat{y}$ as follows:

$$\hat{y}_i(t_i + v h_i) = \sum_{l=1}^{m} L_l(v)\hat{y}_{il},$$

whereby $\hat{y}_{il} = \hat{y}_i(t_i + c_l h_i)$, and in $\sigma_k := (t_k, t_{kj}]$ we have

$$\hat{y}_k(t_k + c_j v h_k) = \sum_{l=1}^{m} L_l(v)\hat{y}_k(t_k + c_j c_l h_k) = \sum_{l=1}^{m} L_l(v)\sum_{s=1}^{m} L_s(c_j c_l)\hat{y}_{ks}.$$

Equation (3.6) now reads

$$\hat{y}_{kj} = c(t_{kj}) + h_k^{1-\alpha} \sum_{l=1}^{m} w_{jl}(\alpha) b(t_{kj}, t_k + c_j c_l h_k, \sum_{s=1}^{m} L_s(c_j c_l) \hat{y}_{ks})$$

$$+ \sum_{i=0}^{k-1} h_i^{1-\alpha} \sum_{l=1}^{m} w_{jl}^{(k,i)}(\alpha) b(t_{kj}, t_{il}, \hat{y}_{il}). \qquad (3.9)$$

In order to obtain $\hat{y} \in S_{m-1}^{(-1)}(Z_N)$, equation (3.9) delivering m, simultaneous equations, (j=1,..,m), will be solved to obtain $\hat{y}_{0,1}, ..., \hat{y}_{0,m}; ...; \hat{y}_{N-1,1}, ..., \hat{y}_{N-1,m}$.

After obtaining $\hat{y}_{kj} = \hat{y}_k(t_{kj})$, we then compute $\hat{y} \in \sigma_k$ as follows:

$$\hat{y}_k(t_k + vh_k) = \sum_{l=1}^{m} L_l(v) \hat{y}_{kl} \text{ with } v \in [0,1], \ (t_k + vh_k) \in \sigma_k,$$

where $L_l(v)$ denotes the lth Lagrange fundamental polynomial for the m collocation parameters.

4.   Order of Convergence

In collocation methods approximating in polynomial spline spaces, the mesh type plays an important role in the order of attainable (global) convergence. Details of the convergence rates and proofs thereto can be seen in the works of Brunner (1985), Brunner and van Houwen (1986) as well as in the article published by Te Riele (1982). In order to simplify our forthcoming arguments and propositions to the convergence rates, we shall illustrate some mesh structures in our working interval $I := [0, T]$:

As a reminder $\prod_N := (0 = t_0 < t_1 < ... < t_N = T)$, and with $\prod_N$ we associate $h_k = t_{k+1} - t_k$ the kth mesh diameter.

a) *Uniform meshes*: A mesh sequence $\prod_N$ is called uniform if

$$t_k := \left(\frac{k}{N}\right) T \quad \text{and} \quad h_k = TN^{-1}, \ (k = 0, 1, ..., N).$$

b) *Asymmetrically graded meshes*: A mesh sequence Π_N is called asymmetrically graded if its grid points are given as follows:

$$t_k := \left(\frac{k}{N}\right)^r T \qquad (k = 0,1,...,N) \text{ and } r \geq 1,$$

whereby the grading exponent r is a real number $(r \geq 1)$. For $r=1$ we obtain the uniform mesh type.

We define the norms applied in the theorems indicated below as follows:

$$\|f\|_\infty := \max\{|f(t)| : t \in [0T]\}.$$

Let x be the exact solution of the problem (2.1) and (2.2) respectively and let $\hat{y} \in S_{m-1}^{(-1)}$ be the approximate discrete collocation solution. We now quote two convergence theorems from Brunner (1985) p. 420, taking into consideration the linear counterpart of (2.2), that is, the function $b(t,s,x)$ is repalced by $b(t,s)x$.

Theorem 4.1: *Let $c \in C^m(I)$ and $b \in C^m(S)$, $S := \{(t,s): 0 \leq s \leq t \leq T\}$ with $m \geq 1$, and assume that neither function vanishes identically, then*

$$\|x - \hat{y}\|_\infty = O(N^{-(1-\alpha)}),$$

for equidistant meshes. The exponent $(1 - \alpha)$ is the best possible convergence order for all $m \geq 1$.

Theorem 4.2: *Let $c \in C^m(I)$ and $b \in C^m(S)$, $S := \{(t,s): 0 \leq s \leq t \leq T\}$ with $m \geq 1$, and assume that neither function vanishes identically. If asymmetrically graded meshes corresponding to the grading exponent $r = \dfrac{m}{1-\alpha}$ are applied, then*

$$\|x - \hat{y}\|_\infty = O(N^{-m}).$$

Remarks: The application of equidistant mesh sequence leads to higher order of convergence of the solution at the right end $t = T$ of the interval $[0, T]$, with an order close to $O(N^{-m})$ (local convergence near T).

With the application of the asymmetrically graded sequence one obtains a better approximation near $t = 0$. It is worth mentioning that asymmetrically graded meshes may however lead to some disadvantages. For instance, due to the fact that the collocation solution is computed recursively starting from a subinterval σ_0 of length $h_0 = TN^{-r}$ with $r = m/(1 - \alpha)$, large values of N will yield very small mesh diameters h_0 and rounding errors may contaminate the subsequent recursive process. This can be demonstrated with the help of the following example:

Setting $\alpha = 0.5$ and $T = 1$ we obtain $h_0 = N^{-2m}$. For the cubic splines (m=4) we obtain $h_0 = N^{-8}$, and for N large enough the initial approximated solutions at the beginning of the interval could be computed without precision.

To avoid this it would be advisable to compute without precision.

The choice of the grading exponent $r= m/(1-\alpha)$ leads to an optimal global convergence of the method. This agrees with the results in approximation theory which states that $O(N^{-m})$ - convergence is the best possible when approximating functions $x(.) \in C^m(I)$ in $S_{m-1}^{(-1)}(Z_N)$ or $S_{m-1}^{(0)}(Z_N)$, Brunner (1985).

5.   Numerical examples

Numerical computation was carried out for examples of Volterra integral equations of the type (2.2):

$$x(t) = c(t) + \int_0^t (t-s)^{-\alpha} b(t,s,x(s))ds.$$

For simplicity in our calculations we set, in all our examples, m=2. We shall also attribute the following values to the collocation parameters c_j for $j = 1,2$:

$$c_1 = (3 - \sqrt{3})/6; \quad c_2 = (3 + \sqrt{3})/6.$$

These points are known as the Gauss-points, in the quadrature method. We have considered the numerical solution of (2.2) with various choices of the function $c(t)$ and different singular kernels, for $t \in [0, T]$. Our computations are also based on the case where $b(t, s, x(s)) = x(s)$.

Example 1: We consider the case of the singular kernel $k(t,s) = (t - s)^{-\alpha}$ and compute the solution of the linear Volterra integral equation

$$x(t) = c(t) + \int_0^t (t-s)^{-\alpha} x(s)ds.$$

The one step discretization (see equation (3.2)) reads

$$x(t) = c(t) + \sum_{i=0}^{k-1} \int_{t_i}^{t_{i+1}} (t-s)^{-\alpha} x(s)ds + \int_{t_i}^{t} (t-s)^{-\alpha} x(s)ds \quad \text{for} \quad k = 0,...,N-1.$$

Setting $t = t_{kj} = t_k + c_j h_k$ and substituting $s = t_i + vh_i$ as well as $s = t_k + vh_k$ in the respective intervals, and a further discretization of the equation leads to the following equations:

$$\hat{y}_{kj} = c(t_{kj}) + h_k^{1-\alpha} \sum_{l=1}^{2} w_{jl}(\alpha) \sum_{s=1}^{2} L_s(c_j c_l) \hat{y}_{ks} + \sum_{i=0}^{k-1} h_i^{1-\alpha} \sum_{l=1}^{2} w_{jl}^{(k,i)}(\alpha) \hat{y}_{il}, \quad j=1,2, \qquad (4.1)$$

whereby $w_{jl}^{(k,i)}(\alpha) = \int_0^1 \left(\frac{t_{kj}-t_i}{h_i} - v\right)^{-\alpha} L_l(v)dv$ with $L_l(v) = \prod_{r=1}^{2}(\frac{v-c_r}{c_l-c_r})$ (r≠ l);

$$w_{j1}(\alpha) = \int_0^{c_j}(c_j - v)^{-\alpha} L_{j1}(v)dv \quad \text{with} \quad L_{j1}(v) = \prod_{r=1}^{2}\left(\frac{v-c_j c_r}{c_j c_l - c_j c_r}\right) \text{ (r≠ l)}. \qquad (4.2)$$

From (4.1) and computing the weights for $l = 1,2$ we obtain for $j = 1, 2$ the following simultaneous equations:

$$\hat{y}_{k1} - h_k^{1-\alpha}\left\{w_{11}(\alpha)\left[L_1(c_1 c_1)\hat{y}_{k1} + L_2(c_1 c_1)\hat{y}_{k2}\right] + w_{12}(\alpha)\left[L_1(c_1 c_2)\hat{y}_{k1} + L_2(c_1 c_2)\hat{y}_{k2}\right]\right\}$$

$$= c(t_{k1}) + \sum_{i=0}^{k-1} h_i^{1-\alpha} \left[w_{1,1}^{(k,i)}(\alpha)\hat{y}_{i1} + w_{1,2}^{(k,i)}(\alpha)\hat{y}_{i2} \right] \qquad (4.3)$$

$$\hat{y}_{k2} - h_k^{1-\alpha} \left\{ w_{21}(\alpha)\left[L_1(c_2 c_1)\hat{y}_{k1} + L_2(c_2 c_1)\hat{y}_{k2} \right] + w_{22}(\alpha)\left[L_1(c_2 c_2)\hat{y}_{k1} + L_2(c_2 c_2)\hat{y}_{k2} \right] \right\}$$

$$= c(t_{k2}) + \sum_{i=0}^{k-1} h_i^{1-\alpha} \left[w_{2,1}^{(k,i)}(\alpha)\hat{y}_{i1} + w_{2,2}^{(k,i)}(\alpha)\hat{y}_{i2} \right] \qquad (4.4)$$

Equations (4.3) and (4.4) can be written together in a compact form as

$$(I - h_k^{1-\alpha} W)\, y = d,$$

whereby

$$I = \text{unit matrix}; \quad y = (\hat{y}_{k1}, \hat{y}_{k2})^T;$$

$$d = \left(c(t_{k1}) + \sum_{i=0}^{k-1} h_i^{1-\alpha}[\ldots], c(t_{k2}) + \sum_{i=0}^{k-1} h_i^{1-\alpha}[\ldots] \right)^T;$$

$$W = \begin{pmatrix} w_{11} L_1(c_1 c_1) + w_{12} L_1(c_1 c_2) & w_{11} L_2(c_1 c_1) + w_{12} L_2(c_1 c_2) \\ w_{21} L_1(c_2 c_1) + w_{22} L_1(c_2 c_2) & w_{21} L_2(c_2 c_1) + w_{22} L_2(c_2 c_2) \end{pmatrix}.$$

We then compute y from

$$y = Q^{-1}d, \quad \text{with } (I - h_k^{1-\alpha}) = Q, \text{ provided the inverse of the matrix } Q \text{ exixts.}$$

This example was tested for $\alpha = 0.5$, $T = 1$, $N = 15$ and the following $c(t)$functions:

a) $c(t) = 1 - 2t^{\frac{1}{2}}$,

Our equation now reads

$$x(t) = 1 - 2\sqrt{t} + \int_0^t (t-s)^{-\alpha} x(s)\,ds.$$

A rough analytic calculation of this Volterra integral equation delivers the solution $x(t) = 1$ for

all $t \in [0, T]$:

Let $t = 0$, $\Rightarrow$ $x(0) = x_0 = c(0) = 1$;

Let $t = 0.5 \Rightarrow x(0.5) = x_1 = c(0.5) + \int_0^{0.5} 0.5(0.5 - s)^{0.5} x_0 \, ds$

$$x_1 = 1 - 2\sqrt{0.5} - 2x_0\{\sqrt{0.5 - s}\}_0^{0.5} = 1.$$

A numerical computation of Volterra integral equation with graded mesh sequence delivers the result in Table 1.

t	Exact Solution $x(t)$	Approximated Solution $\widehat{x}(t)$	Relative Error
0.000158	1.00000	1.00000000000	0.000000
0.000257	1.00000	1.00000000000	0.000000
0.001343	1.00000	1.00000000000	3.64 E-12
0.004365	1.00000	0.99999999997	3.09 E-11
0.010888	1.00000	0.99999999998	1.82 E-11
0.022949	1.00000	1.00000000080	7.64 E-10
0.043062	1.00000	1.00000000130	1.29 E-09
0.074212	1.00000	0.99999999974	1.26 E-09
0.119862	1.00000	0.99999999855	1.45 E-09
0.183945	1.00000	1.00000001250	1.45 E-08
0.270870	1.00000	0.99999999832	1.68 E-09
0.385521	1.00000	1.00000001000	9.99 E-09
0.533254	1.00000	1.00000015720	1.53 E-07
0.719901	1.00000	1.00000022010	2.20 E-07
0.951767	1.00000	1.00000050990	5.10 E-07

Table 1 $: c(t) = 1 - 2t^{\frac{1}{2}}$

Example 2: This example handles a Volterra integral equation with the following weakly singular kernel:

$$k(t,s) = 1 + 2\sum_{n=1}^{\infty} e^{-n^2\pi^2(t-s)}.$$

This kernel originates from the weak solution of the heat flow parabolic differential equation. The Volterra integral equation thereto reads:

$$x(t) = c(t) - \int_0^t \left(1 + 2\sum_{n=1}^{\infty} e^{-n^2\pi^2(t-s)}\right) x(s) \tag{4.5}$$

Analogue to example 1, equation (4.5) after discretization and approximation will deliver

$$\hat{y}_{kj} = c(t_{kj}) + \sum_{i=0}^{k-1} h_i \sum_{l=1}^{2} \int_0^1 \{1 + 2\sum_{n=1}^{\infty} e^{-n^2\pi^2(t_{kj}-t_i-vh_i)}\} L_l(v)\hat{y}_{il}\,dv$$

$$+ h_k \sum_{l=1}^{2} \int_0^{c_j} \{1 + 2\sum_{n=1}^{\infty} e^{-n^2\pi^2(c_j-v)h_k)}\} L_{jl}(v) \sum_{s=1}^{2} L_s(c_j c_l)\hat{y}_{ks}\,dv \; ; \quad j=1,2 \qquad (4.6)$$

The weights of this equation are

$$w_{jl}^{(k,i)} = \int_0^1 \{1 + 2\sum_{n=1}^{\infty} e^{-n^2\pi^2(t_{kj}-t_i-vh_i)}\} L_l(v)\,dv,$$

$$w_{jl} = \int_0^{c_j} \{1 + 2\sum_{n=1}^{\infty} e^{-n^2\pi^2(c_j-v)h_k)}\} L_{jl}(v)\,dv \qquad \text{for j=1,2 and l=1,2.}$$

These deliver, after application of the partial integration method and substitution of the integration limits, and analogous to equations (4.3) and (4.5) the following in a compact form

$$(I - h_k^{1-\alpha} W)\, y = d,$$

Remarks: The computation of the infinite sequences occurring in the quadrature weights shall be approximated as follows: For example

$$\sum_{n=1}^{\infty} \frac{1}{n^\beta} e^{-n^2\varepsilon} = \sum_{n=1}^{r} \frac{1}{n^\beta} e^{-n^2\varepsilon} + \mathrm{Re}\,m. \qquad (4.7)$$

Our aim here is to sum the sequence in (4.7) up to a certain integer, r, such that the remainder term, *Rem,* will be negligible.

We now compute (4.5) for two different $c(t)$– functions. We set $N = 15$, and $T = 1$:

a) $\quad c(t) = 1 - t - 2\sum_{n=1}^{\infty} \frac{2}{n^2\pi^2}(1 - e^{-r^2\pi^2 t}),$

The exact solution of the Volterra integral equation thereto is $x(t) = 1$ for all $t \in [0,T]$, (See Table 2a).

t	Exact Solution $x(t)$	Approximated Solution $\widehat{x}(t)$	Relative Error
0.000158	1.0000000	1.00000000010	1.02 E-09
0.000257	1.0000000	1.00000000976	2.38 E-09
0.001343	1.0000000	1.00000000020	2.04 E-09
0.004365	1.0000000	1.00000000011	1.13 E-09
0.010888	1.0000000	0.99999999978	2.16 E-09
0.022949	1.0000000	1.00000000011	1.11 E-09
0.043062	1.0000000	1.00000000009	8.53 E-10
0.074212	1.0000000	0.99999999992	8.37 E-10
0.119862	1.0000000	1.00000000004	3.67 E-10
0.183945	1.0000000	1.00000000003	3.22 E-10
0.270870	1.0000000	1.00000000001	8.37 E-11
0.385521	1.0000000	1.00000000001	1.13 E-10
0.533254	1.0000000	1.00000000002	1.53 E-10
0.719901	1.0000000	1.00000000002	2.18 E-10
0.951767	1.0000000	1.00000000003	2.95 E-10

Table 2a: $c(t) = 1 - t - 2 \sum_{n=1}^{\infty} \frac{2}{n^2 \pi^2} (1 - e^{-r^2 \pi^2 t})$,

b) $c(t) = t - \dfrac{t^2}{2} - \dfrac{2t}{\pi^2} 2 \sum_{n=1}^{\infty} \frac{1}{n^2} + \dfrac{2}{\pi^4} \sum_{n=1}^{\infty} \frac{1}{n^4} (1 - e^{-r^2 \pi^2 t})$.

The exact solution of the Volterra integral equation thereto is $x(t) = t$ for all $t \in [0,T]$, (see Table 2b).

t	Exact Solution $x(t)$	Approximated Solution $\widehat{x}(t)$	Relative Error
0.000158	0.000158	0.0000158025	9.27 E-15
0.000257	0.000257	0.0002567901	2.49 E-14
0.001343	0.001343	0.0013432099	7.11 E-14
0.004365	0.004365	0.0043654321	2.13 E-14
0.010888	0.010888	0.0108879012	2.84 E-14
0.022949	0.022949	0.0229491358	2.84 E-14
0.043062	0.043062	0.0430617284	5.68 E-14
0.074212	0.074212	0.0742123457	2.27 E-13
0.119862	0.119862	0.1198617284	1.14 E-13
0.183945	0.183945	0.1839446914	2.27 E-13
0.270870	0.270870	0.2780701235	4.55 E-13
0.385521	0.385521	0.3855209877	1.36 E-12
0.533254	0.533254	0.5332543210	1.82 E-12
0.719901	0.719901	0.7199012346	1.82 E-12
0.951767	0.951767	0.9517669135	9.10 E-13

Table 2b: $\quad c(t) = t - \dfrac{t^2}{2} - \dfrac{2t}{\pi^2} 2\sum_{n=1}^{\infty} \dfrac{1}{n^2} + \dfrac{2}{\pi^4} \sum_{n=1}^{\infty} \dfrac{1}{n^4}(1 - e^{-r^2\pi^2 t}).$

6.   References

Brunner, H. (1985). The numerical solution of weakly singular Volterra equations by

collocation on graded meshes. *Mathematics of Computation Vol. 45 No. 172*, pp 417-437.

Brunner, H.   The history of numerical methods for Volterra integral equations.

Institut Université de Fribourg, Switzerland.

Brunner, H. & van der Houwen, P. J. (1986). The numerical solution of Volterra equations.

CWI Monographs 3 (Centrum voor Wiskunde en Computer Science) North Holland.

Brunner, H., Pedas, A. & Vainikko, G. (1997). The piecewise polynomial collocation method

for nonlinear weakly singular Volterra equations. *Helsinki University of Technology,*

Institute of Mathematics, Research Reports A392.

Diogo, T. & Lima, P. (2007). Collocation solution of weakly singular Volterra integral equation.

TEMA Tend. Mat. Apl. Comput. 8 No. 2, 229-238.

Gohberg, I., & Goldberg, S. (1981). *Basic Operator Theory.* Birkhuser.

Jalalvand, M., Jazbi, B, & Mokhtarzadeh, M. R. (2013). A finite difference method for

the smooth solution of linear Volterra integral equations. *Int. J. Nonlinear Anal. Appl. 4*

No. 2, 1-10.

Linz, P. (1985). Analytical and numerical methods for Volterra equations. *SIAM*

Philadelphia

Ljusternik, L. A., & Sobolew, W. I. (1976). *Elements of functional analysis.* Akademie –

Verlag, Berlin.

Makroglou, A., (1981, July). A block-by-block method for Volterra integro-differential

equations with weakly singular kernels. *Mathematics of Computation Vol. 37 Number*

155, pp 95-99.

Te Riele, H. J. J., (1982). Collocation methods for weakly singular second kind Volterra integral equations with non-smooth solution. *IMA Jounal of Numerical Analysis* 2, 437 – 449.

YOUR KNOWLEDGE HAS VALUE

- We will publish your bachelor's and
  master's thesis, essays and papers

- Your own eBook and book -
  sold worldwide in all relevant shops

- Earn money with each sale

Upload your text at www.GRIN.com
and publish for free